MEDIA

NADIA L. HOHN

SERIES EDITOR • TOM HENDERSON

www.rubiconpublishing.com

Associate Publisher: Amy Land
Project Editor: Jessica Rose
Editor: Kaitlin Tremblay
Creative Director: Jennifer Drew
Lead Designer: Sherwin Flores
Graphic Designers: Jen Harvey, Megan Little, Jason Mitchell

15 16 17 18 19   5 4 3 2 1

ISBN: 978-1-77058-834-9

Printed in China

CONTENTS

4 **Introduction: Media**

6 **Sidekicks and Stereotypes**
How well are people from racialized groups represented in the media? Read this infographic to find out.

8 **Film Festivals Around the World**
These interesting fact cards will introduce you to film festivals around the world.

10 **Bino and Fino: A Nigerian Cartoon Brand for Kids**
In this interview, learn all about why Adamu Waziri made his own cartoon about life in Nigeria.

14 **Kickin' It: Running Shoes and the Evolution of Influence**
Running shoes tell stories just like any other form of media. Learn more in this report.

18 **Wanting to Be Heard**
Mary Ann Shadd was the first woman newspaper publisher in Canada. Learn why her achievements matter in this profile.

22 **Crafting the Code**
What would it be like to be sent back to the time of the Underground Railroad? That's exactly what happens in this short story.

28 **Speaking Out and Stepping Up**
Find out what some of the media world's most creative minds have to say about the topics of representation, inspiration, and more in this collection of quotations.

31 **Mom Gives Barbies a Multicultural Makeover**
Why do you think a mother made diverse Barbie dolls for her daughter? Find out in this online article.

34 **What Do You Want to Be?**
There are many different jobs related to the media. Read these fact cards to learn about a few different ones.

38 **These T.V. Men and Women**
In this poem, Maxine Tynes talks about the people on television who have made a difference in her life.

40 **Media Whiz Kids**
Read these profiles to learn about incredible kids who are doing amazing things with media.

42 **People of All Shades**
What is shadeism? Find out what it is and why it's dangerous in this report.

44 **Afrofuturism**
Artists use science fiction to share their unique experiences. The result is Afrofuturism. Learn more in these profiles.

48 **Acknowledgements**

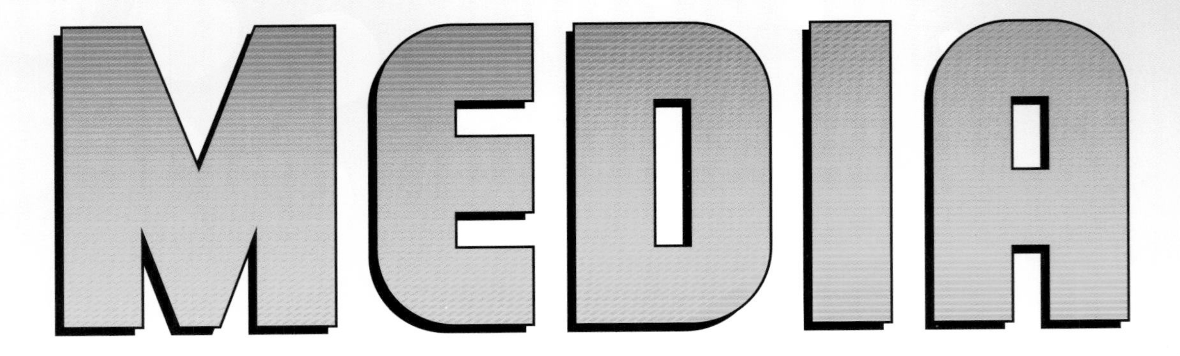

Have you ever thought about your relationship with the media? If you're like the average young person, you may be spending up to eight hours a day with various forms of media. Whether you're playing computer or video games, watching TV, or on the Internet or your phone, all this time with media shapes the way you see the world.

It is important to remember that all media messages have been created purposefully to achieve a specific goal. Regardless of whether the goal is to entertain, educate, sell products, sway beliefs, or change the world, it is important to view media critically. This can be done by asking yourself who created the media product and why. It is important to look for biases and stereotypes. Remember also to pay attention to whose story is being told and whose is not.

biases: *beliefs that certain people or ideas are better than others*
stereotypes: *oversimplified impressions of a person based on an aspect of his or her identity*

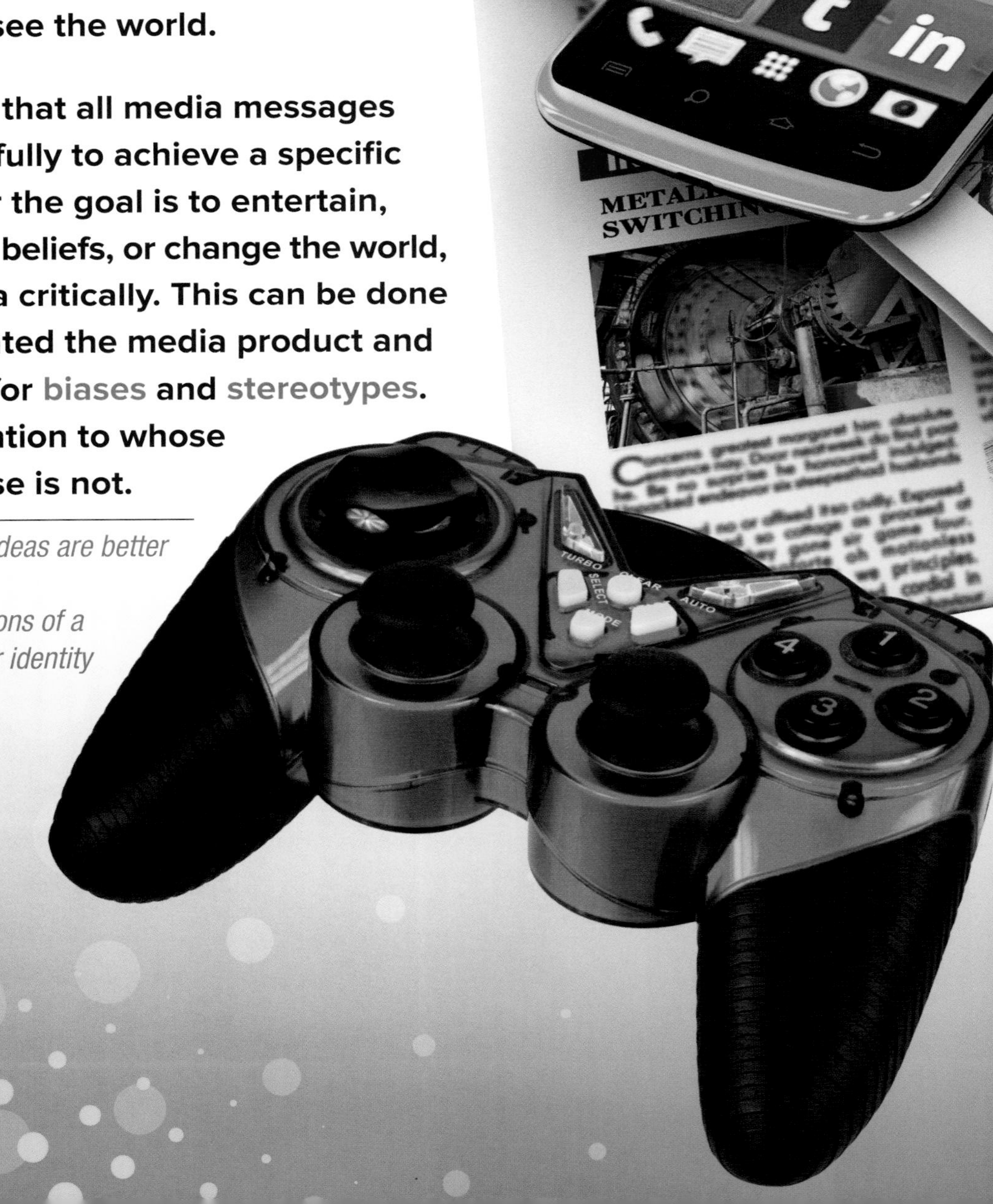

How are people of African descent changing the media landscape?

SIDEKICKS AND STEREOTYPES

THINK ABOUT IT

What stereotypes exist about people within your school, your community, and Canada?

ACCORDING TO MEDIA SMARTS, Canada's Centre for Digital and Media Literacy, racialized people in Canada are under-represented or misrepresented in the media. Representations in the media influence how people see and value themselves and others. Read this infographic to learn more about this important issue.

racialized: *given a racial identity that is neither Caucasian nor Aboriginal*

In 2011, **one** out of every **five** people in Canada identified themselves as a member of a group considered to be racially visible. This equalled **6 264 800 people.**

In 2010 and 2011, out of **14 original television programs** airing on CBC, CTV, and Global, **only one** had a lead character from a racialized group. This show was *Little Mosque on the Prairie.*

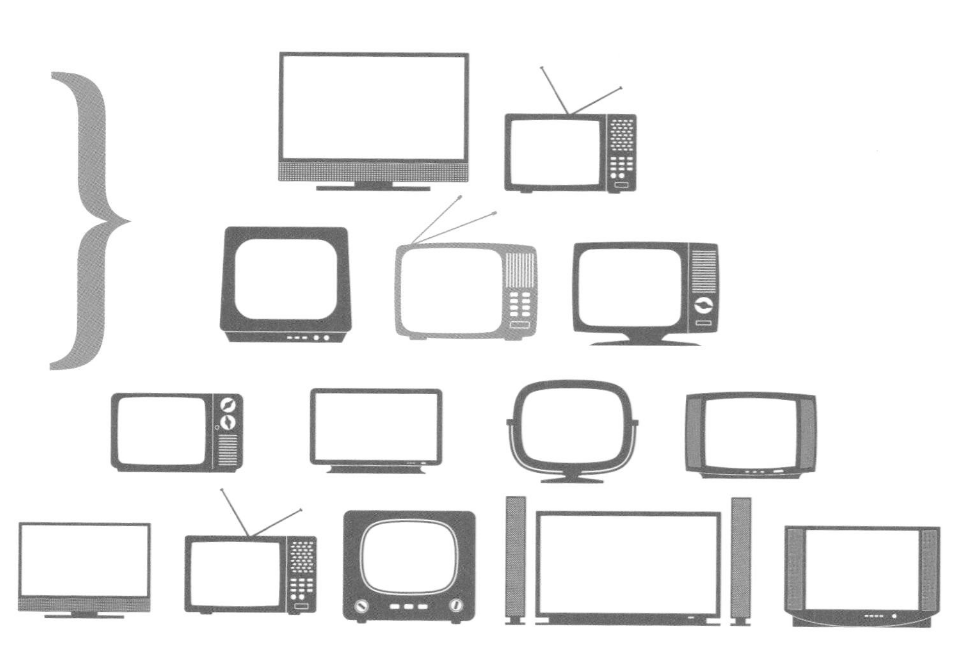

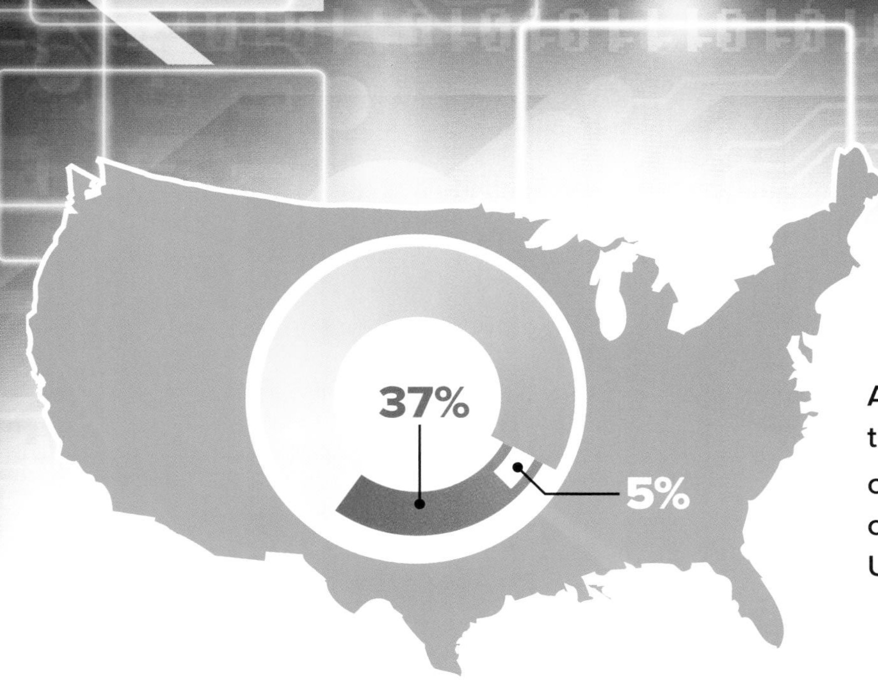

About **37 percent** of the population of the United States is made up of people of colour. However, only **five percent** of children's books published annually in the United States are written by people of colour.

In 2009, a study of video game advertising found that only **23 percent** of characters were from a racialized group.

The same 2009 study found that **100 percent** of Black men in video games were portrayed as stereotypically violent or athletic, or both.

According to a study by the University of Wisconsin, in the United States, only **93** of the **3200** children's books published in 2013 were about Black people.

Only **four percent** of female characters and **12 percent** of male characters in Canadian television dramas are from racialized groups.

CONNECT IT

Pick a favourite television show, book, or video game. How much diversity is represented in it? Summarize your findings to a classmate, discussing whether there is enough diversity in the example you chose.

Film Festivals Around the World

THINK ABOUT IT

How do you decide which films you are going to see? Write a list of criteria, ranking each item from most important to least important.

CHANCES ARE YOU like going to see movies on the big screen. But have you ever been to a film festival? Film festivals take place all over the world. Some specialize in films made by and starring Black people in the movie industry. Read all about some of these film festivals in these fact cards.

BLACK INTERNATIONAL CINEMA BERLIN

LOCATION: Berlin, Germany

FOUNDED: 1986

FEATURES: Films that reflect the experiences of people living in Africa and in the African diaspora, as well as films by other filmmakers, regardless of race, religion, or gender

WHY? Black International Cinema also showcases films that reflect other cultures. It aims to foster greater respect for all cultures, while educating people about a variety of issues affecting people of African descent. The festival includes film presentations, seminars, discussions, and exhibitions.

IMAGES OF BLACK WOMEN FILM FESTIVAL

LOCATION: London, England

FOUNDED: 2004

FEATURES: Films from women of African descent

WHY? Images of Black Women Film Festival (IBW) celebrates and promotes women of African descent in cinema, whether they are onscreen or behind the camera. IBW promotes race and gender equality in films in order to encourage a new generation of filmmakers. Other goals of IBW include increasing the visibility of women of African descent in film, showcasing ways for more women to get involved in film, and acknowledging the roles of Black women in film. IBW also hosts workshops and seminars, including an animation workshop for eight- to 14-year-old children.

Volunteers teach children about film production.

LOLA KENYA SCREEN

LOCATION: Nairobi, Kenya

FOUNDED: 2005

FEATURES: An audiovisual media festival and mentorship program for children and youth

WHY? Lola Kenya Screen is a festival that is exclusively created for children and youth. According to Lola Kenya Screen, "Children are agents of change." The festival promotes talent among children and youth through hands-on skill development and programs in various forms of media, such as journalism and filmmaking, and also runs a film festival. Lola Kenya Screen teaches children and youth how to make multimedia that support issues such as literacy, gender equality, self-expression, and democracy.

mentorship: *teaching and guiding*

MONTREAL INTERNATIONAL BLACK FILM FESTIVAL

LOCATION: Montreal, Quebec

FOUNDED: 2005

FEATURES: Independent films that showcase the different realities Black people face around the world

WHY? The Montreal International Black Film Festival (MIBFF) promotes the independent film industry, showcasing films about Black people from around the world. The MIBFF is also a place where people can discuss cultural issues that affect their local and global communities. These discussions are designed to help people get to know one another better and avoid misunderstandings. The MIBFF was the Montreal Haitian Film Festival before it transitioned in 2010 to focus on films from around the world. The festival is bilingual, showing films in both French and English.

CONNECT IT

Imagine you are creating your own film festival to highlight films that are under-represented in the mainstream media. Choose a theme and a name for your film festival, and write one sentence that describes your film festival's purpose.

Bino and Fino: Cartoon Brand

THINK ABOUT IT

You've probably heard about the negative effects of television on children. Television shows can have a positive impact, too. Tell a classmate about one television show that has had a positive impact on you.

Adamu Waziri

AS A CHILD growing up in Nigeria, animator Adamu Waziri watched foreign cartoons. These cartoons showed a life very different from the one he knew in Nigeria. He decided to do something about that. He created a cartoon that reflected the lives of children in Nigeria. His cartoon features a brother and sister, named Bino and Fino, who live in Nigeria with their grandparents, Mama Mama and Papa Papa. Read the following interview to learn more about why and how Waziri made *Bino and Fino*.

A Nigerian for Kids

Fino

HELEN NGOH
VENTURE CAPITAL FOR AFRICA
26 MARCH 2012

Helen Ngoh: [Tell] us more about *Bino and Fino.*

Adamu Waziri: *Bino and Fino* is a Nigerian educational cartoon brand aimed at three- to five-year-olds. We are working to identify business models that would allow African animation to thrive as a business. …

HN: Why did you come up with this business?

AW: I decided to start the *Bino and Fino* brand from a cultural and practical standpoint. Children love cartoons. However, most cartoons shown in Nigeria and other parts of Africa are foreign and imported. I felt there was a need to produce local, animated, educational content to cater [to] Nigerian and African families. …

HN: At what [stage] is *Bino and Fino*?

AW: The first pilot episodes, or TV features, have been produced. Those are now being aired in the United Kingdom (UK) and South Africa. We are currently working on getting it aired on Nigerian TV. The DVD is on sale in certain bookshops and online via the cartoon's website. We are now talking to investors and sponsorship partners who see the value and potential of the brand. We are preparing to start producing the first season of 13 episodes.

Since the first 13 episodes were released on DVD, Waziri started producing a full season of 26 episodes.

HN: What kind of team is working on this venture?

AW: The team producing the show consists of three animators on the production side. The rest of the creative talent, such as writers and actors, is outsourced to freelancers. …

HN: What's the space like at the moment? Do you have competition?

AW: In Nigeria, there isn't much local indigenous competition in the market. We have first-mover advantage, though there are a few other projects in development, so the scenario is likely to change. As far I know, *Bino and Fino* is the only Nigerian educational cartoon out there.

"First-mover advantage" means that there is very little competition for those who get a product into the market before anybody else. Do you think it is a good thing that there are other projects in development?

venture: *business, one that usually involves some risk*
freelancers: *people who work for different companies on contract*
indigenous: *originating from within a place or culture*

Mama Mama, Fino, Bino, and Papa Papa

HN: What is the level of African animation at the moment, compared maybe to the US or European industry?

AW: Well, the industry is in its infancy. It's not close to the US, European, or Asian market. I'd say the most-developed countries are South Africa, Kenya, Ethiopia, and Egypt to my knowledge. But it's a fast-moving industry, so things are constantly changing. Obviously, the talent and passion are there. The surrounding structures to enable the industries to grow are the problem.

Zeena the Magic Butterfly

HN: Tell us about your challenges and achievements with this venture.

AW: Well, we are the first company to actually produce a long-form children's education program in Nigeria and get it to market globally. We've developed techniques that mean we can complete high-standard animations in the Nigerian operating environment.

The cartoon now has a group of passionate fans from around the world who love the show and want more. Thanks to online broadcast and TV airing in the UK on Sky TV and South Africa, that following is growing. The project is now getting increased media coverage by organizations such as CNN. …

Why is the Internet important for people like Waziri who want to create media for a specific audience?

MEET KAMI

In South Africa, *Takalani Sesame* created a Muppet named Kami. Kami is a five-year-old who is HIV positive. Kami helps fight the negative beliefs associated with HIV/AIDS, and teaches important facts about the disease. Kami often plays the "train game" with others, where kids join together and form a train. This is important because it shows that Kami's friends include her despite her having HIV.

Kami

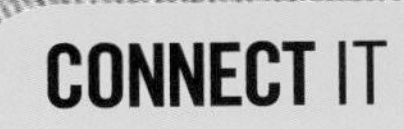

Create a storyboard for an episode of your favourite show. Be sure your storyboard for the show reflects the reality and diversity of the world. Present your storyboard to your classmates, and answer any questions they have.

KICKIN' IT:
RUNNING SHOES AND THE EVOLUTION OF INFLUENCE

THINK ABOUT IT

Who or what influences your decisions about what you wear?

MOST OF US are surrounded by media messages on a daily basis. These messages can affect the way we think about everything from happiness to beauty. They can even affect how we behave. Often, we don't realize how powerful these messages are. Sure, some messages, such as "Buy now! Limited time only!" are easy to notice. Others, however, are not as obvious.

Have you ever thought about what it is that makes someone want to pay $160 for a pair of Air Jordans instead of $20 for some no-name shoes? These reports show the various ways shoe companies have successfully used sports and music stars to market their products.

Jesse Owens and Adidas

Around 1920, German shoemaker Adolf "Adi" Dassler and his brother started making athletic shoes. At this time, there weren't any mass media advertising campaigns as we know them today. So, to help spread the word about their innovative footwear, they gave free shoes to top athletes.

This strategy really paid off when African American track star Jesse Owens wore their shoes to the 1936 Olympics in Berlin, Germany, where he won four gold medals. At this time in history, Adolf Hitler was ruling Germany and promoting racist ideas about Aryan superiority. He wanted the 1936 Olympics to showcase Aryan superiority, but Owens's performance stole the show. Photos of Owens in Dassler's shoes appeared in newspapers around the world. After parting ways with his brother, Dassler created his own shoe company. The company was called Adidas.

Aryan: *Caucasian not of Jewish descent; associated with the Nazis*

Jesse Owens

Adi Dassler's first athletic shoe

Michael Jordan and Nike

In 1985, Michael Jordan was a young basketball star playing for the Chicago Bulls. At the time, players had to wear basketball shoes, running shoes designed specifically for playing basketball, to match their jerseys and their teammates. However, this didn't stop Jordan. Nike released a red and black running shoe called the Air Jordan 1 for him to wear that didn't follow the rules set by the National Basketball Association (NBA). By wearing these shoes, Jordan broke the rules, and he was fined $5,000 for every game he played. Nike not only paid the fines each time Jordan wore the shoes, but they made a commercial about it. "On October 15th, Nike created a revolutionary new basketball shoe," said the voice in the ad. "On October 18th, the NBA threw them out of the game. Fortunately, the NBA can't keep you from wearing them."

After this, sales for these basketball shoes increased because they were a sign of rebellion. Jordan rebelled when he wore these shoes, and consumers wanted to wear shoes that had this story behind them. Nike founder Phil Knight attributed the shoes' success to "the perfect combination of quality product, marketing, and athlete endorsement."

endorsement: *praise; sponsorship*

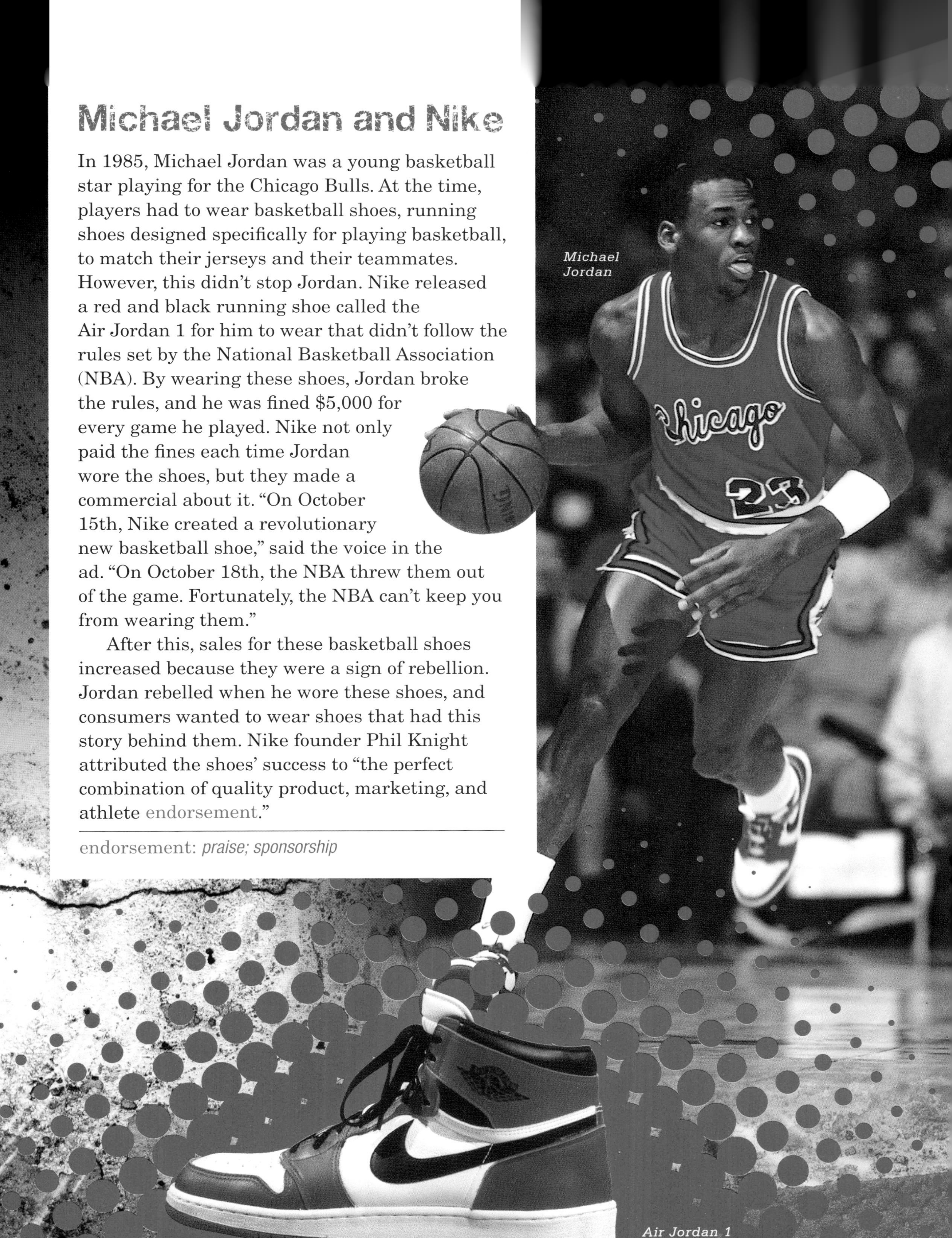

Michael Jordan

Air Jordan 1

Run-DMC

Run-DMC and Adidas

You've probably noticed that popular music lyrics and videos, including rap and hip hop, often mention brand names. In some cases, the artists are even paid to mention products. However, things were much different in the very early days of rap. In 1983, Run-DMC, a rap trio from Queens, New York, was just starting to become popular. They wore track pants, gold chains, and Adidas shoes. They loved the shoes so much that in 1986 they wrote the song "My Adidas." They were not paid by Adidas to write the song.

That same year, at a Run-DMC concert in Madison Square Garden, Run stopped the song and asked everyone to take off a shoe and hold it up. The group's producer had made sure that Adidas executives were at the show so that they could see the influence Run-DMC had on their fans' buying and wearing Adidas. After this, Run-DMC received $1.5 million to endorse the brand. At that time, "the idea that rap artists could land an endorsement deal for a mainstream product was unthinkable," wrote the *Village Voice*. "[Now] rappers are well aware that every brand name they drop can mean money — both officially and under the table."

"We sang 'My Adidas' because we just liked them," DMC said. "That's the difference. Now a lot of guys are just hoping to get that phone call."

In 2013, Canadian rapper Drake teamed up with Nike to design his very own running shoe. In 2014, Drake partnered with Michael Jordan's Nike brand, and he designed both the OVO Air Jordan 10 and 12. Also in 2014, Drake designed the black and gold version of the OVO Air Jordan 6 running shoes. These Air Jordan 6 running shoes originally started off as an April Fool's joke, but it ended up inspiring Drake to make the shoes a reality.

Drake

Vashtie Kola — Influencer

Vashtie Kola has been called a "streetwear tastemaker," a "top sneaker influencer," and a "renaissance woman." Vashtie Kola is a fashion designer, DJ, and video director. She was also Nike's first female Air Jordan designer. As an influencer, she is part of a group of trendsetters who are able to partner with brands to communicate messages about these brands to their loyal social media followers. Companies often hire influencers to reach customers in ways traditional advertisements can't. "It's an easy way for brands to get to the consumer," Kola says. "It's easier for a big corporation to see what their customers are thinking and talking about." Kola knows her influence goes beyond just helping kids decide which brand to buy. Her parents are from Trinidad, and Kola learned to work hard from them. Of her role as an influencer, she said, "Seeing me live this lifestyle could help impact someone else's life who grew up in a small town like I grew up in."

The term "Renaissance woman" refers to someone who is skilled and knowledgeable in various and different fields. The word refers to the Renaissance in Western Europe between the 14th and 16th centuries, where many people made significant advances in art and culture.

Vashtie Kola

CONNECT IT

Based on what you learned in this report, design your own running shoe. As you design your shoe, think about the following questions: Who is your audience? How much should you charge for your shoe? What message do you want to send with your shoe? How will you convey the message?

WANTING TO BE HEARD

THINK ABOUT IT

Where do you prefer to get news? Do you get news from a newspaper, a website, or a television show? Tell a classmate why you prefer this medium.

MARY ANN SHADD was an anti-slavery activist, lawyer, journalist, and publisher. Read the following profile to learn how she became the first woman publisher in Canada.

PROVINCIAL FREEMAN.

First issue of the
Provincial Freeman

On 24 March 1853, Mary Ann Shadd published the first issue of the *Provincial Freeman*, an anti-slavery newspaper in Windsor, Ontario, catering to abolitionists. This made her the first female publisher in Canada, and the first Black female publisher in North America. But Shadd's road to success wasn't easy.

Shadd was born in Wilmington, Delaware, in 1823. According to the United States census, by 1820, there were 1 538 038 enslaved people living in the United States. Even those born free, like Shadd, lived with the fear of being kidnapped and enslaved.

abolitionists: *people fighting to end slavery*

When Shadd was a child, most schools were segregated. Formal education for African American children was difficult to find. Wanting their children to be educated, Shadd's parents, Harriet and Abraham Shadd, moved their family from Delaware to Pennsylvania. This move meant that Shadd and her 12 younger brothers and sisters could go to a good school. It also allowed Shadd's father to continue his involvement with the abolitionist movement. The Shadds hosted guests and even helped to hide fugitive slaves. Their home was a "station" on the Underground Railroad and Abraham Shadd was a "conductor."

Mary Ann Shadd

At the time, a woman publishing a newspaper was unheard of.

Years later, Shadd and her brother Isaac left Pennsylvania to come to Ontario, or Canada West, as it was called then. Her anti-slavery activism continued the legacy started by her parents. Shadd opened a school outside Windsor, Ontario, and became a teacher. At the time, integration was forbidden, so Black people and White people were not allowed to attend the same schools. Shadd fought against this, and made her new school integrated. Fighting against segregation was a way for her to fight against discrimination. Other members of her family would also eventually settle in Ontario.

The *Provincial Freeman* became a way for Shadd to continue to voice her abolitionist messages. At the time, a woman publishing a newspaper was unheard of. Many people were critical of women and thought they were inferior to men. Because of this, Shadd signed articles with her initials, *M.S.* She chose to list the names of Samuel Ringgold Ward and Reverend Alexander McArthur on the newspaper as the editors, even though she was often the editor. She felt she needed to do this to satisfy the newspaper's funders and to silence critics.

Do you agree with Shadd's decision to put the names of men on the front of the paper? Explain your answer.

segregated: *separated based on race*
integration: *bringing people of different races together*

Shadd worked tirelessly, travelling throughout Canada and the United States. She gave public speeches and gained financial support for the newspaper. Even though she often put the names of men on the front page, it was very clear that she wrote the fiery abolitionist messages, which were also a fight for women's rights.

Shadd was outspoken and a strong debater. She often critiqued the work of other abolitionists, including escaped slave Henry Bibb. Shadd thrived at conferences, where she was often one of the few women present.

On 3 January 1856, at the age of 32, Shadd married Thomas Cary. In Cary, she found a true partner and a shared sense of purpose in the anti-slavery movement. Shadd became a stepmother to his children, and then they had two more children together. Despite her hectic life, Shadd continued to fundraise for the newspaper.

She continued to publish the *Provincial Freeman* until around 1860. Around this same time, Cary died at the age of 35. Although this brought an end to one phase of her life and career, Shadd started a new phase. She began recruiting Black soldiers for the American Civil War. At the age of 60, she went on to study law at Harvard University. This made her one of the first Black women to get a law degree in the United States.

Shadd died on 5 June 1893 in Washington, DC, but her legacy continues to live on in her many descendants in North America. There have also been songs, movies, and schools created in her name.

A statue of Mary Ann Shadd in the B.M.E. Freedom Park, Chatham, Ontario

Shadd's Legacy Continues

As in the time of the *Provincial Freeman* and other abolitionist newspapers, the Black press today helps to ensure that all voices are heard on issues that directly affect the African Canadian community. Patricia Bebia Mawa, the co-founder of *Planet Africa Television*, makes sure that there exists media specifically for people of African descent living in Canada. Like Shadd, Patricia Bebia Mawa was persistent in achieving her dream of creating diverse media for African Canadians.

Planet Africa Group produces the television show *Planet Africa Television*. Patricia Bebia Mawa is the host of *Planet Africa Television*, and she is also the managing editor of *Planet Africa Magazine*. In 1994, she immigrated to Canada from Nigeria. Along with her husband, Moses A. Mawa, she co-founded the Planet Africa Group in 2002. Moses A. Mawa states that Planet Africa Group was established to "inspire, educate, and entertain our community." Planet Africa Group aims to showcase positive Black role models for youth.

Planet Africa's parent company, Silvertrust Communications, now produces television shows and documentary films. It also publishes many magazines. Patricia Bebia Mawa started Silvertrust in 1996 while she was studying film at Carleton University in Ottawa, Ontario. *Diversity Magazine* is created by Silvertrust Communications. It focuses on showcasing different cultures. When asked if there will be always a need for this kind of platform, she explained that there will "always be a need for African Canadian newspapers, magazines, films, and other media to address the unique challenges and aspirations of our community." Since coming to Canada, she has won the International Women Achievers' Award and the Toronto Police Community Service Award.

Moses A. Mawa states that Planet Africa Group was established to "inspire, educate, and entertain our community." Planet Africa Group aims to showcase positive Black role models for youth.

Patricia Bebia Mawa with an issue of Diversity Magazine

CONNECT IT

Find other examples of media in Canada that celebrate a group of people or culture. For example, you can learn more about the Aboriginal Peoples Television Network (APTN). Prepare a short report summarizing what you've learned about the example you find most interesting.

Crafting THE CODE

THINK ABOUT IT

With a partner, brainstorm some ways in which the media can be used to help empower individuals or groups.

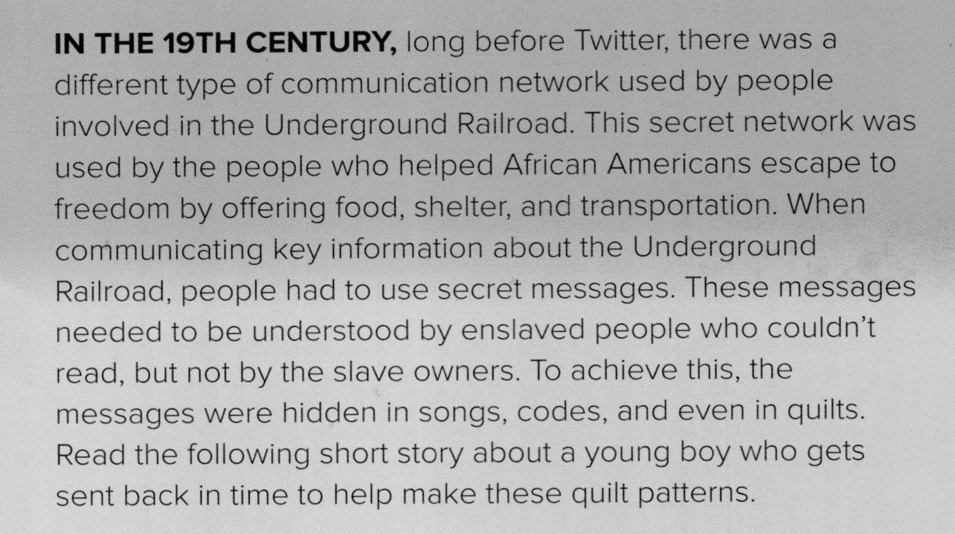

IN THE 19TH CENTURY, long before Twitter, there was a different type of communication network used by people involved in the Underground Railroad. This secret network was used by the people who helped African Americans escape to freedom by offering food, shelter, and transportation. When communicating key information about the Underground Railroad, people had to use secret messages. These messages needed to be understood by enslaved people who couldn't read, but not by the slave owners. To achieve this, the messages were hidden in songs, codes, and even in quilts. Read the following short story about a young boy who gets sent back in time to help make these quilt patterns.

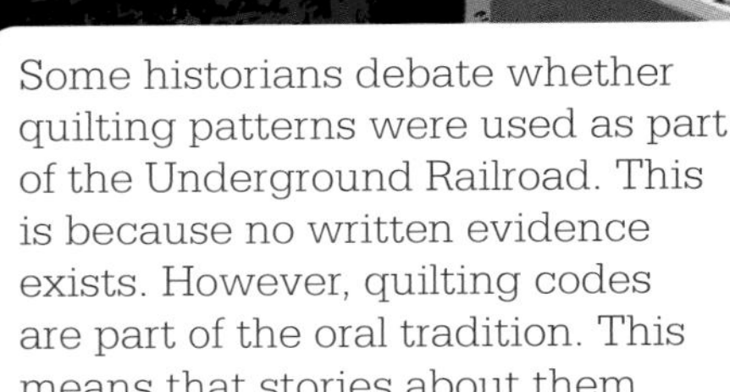

Some historians debate whether quilting patterns were used as part of the Underground Railroad. This is because no written evidence exists. However, quilting codes are part of the oral tradition. This means that stories about them have been passed down orally from generation to generation.

"You've made some bad choices, kid." From across the room, the sergeant glanced at me before turning to my mom. To her, he said, "Thanks for coming to the station, Ms. Powell. Your son here is in a lot of trouble."

"Don't worry, officer. I'll make sure Dwayne learns his lesson." My mom was polite with the sergeant, but I could tell she was furious.

The plan that got me into this mess was flawed from the start. I should have known better. When Tyrell saw the doodles in my book, he asked me to help him draw his "tag" on the side of a train.

"Draw what?" I asked, since I had no clue what he was talking about.

"Don't tell me you don't know what taggin' is?" replied Tyrell.

"Sure, I do," I lied. The words just slid right out before I could ask any more questions.

That night, I met Tyrell by the abandoned warehouse near the train tracks.

"What's the matter? You scared?" he asked.

"Nah ...," I shook my head. "I got this."

I held the can, tilted it, and pressed the button. The *hsssssshhh* was deafening and stained a lime green streak on the boxcar. I began to add baby blue and lemon yellow.

I got so caught up in what I was doing that I didn't hear the sound of Tyrell's feet as he took off. I'm not even sure what happened next exactly. It all seemed like I was in a movie; it was so surreal. I don't remember how I got caught, but the next thing I knew, I was in a police car.

Mom gave me an earful during the tense ride home from the station. When we got home, she was on the phone right away, calling up Uncle Fuzz and Aunt Frederica. My mom didn't play around. This wasn't the first time I had messed up, and she had warned me before that if I did it again, she would send me straight to my aunt and uncle's farm. End of discussion.

"They have plenty of land for fresh air and exercise. And they have lots to do to keep idle hands busy — you won't have time to find trouble," my mom said.

My only memories of my aunt and uncle's place were mosquitoes and black flies in the summer, my annoying little cousins, and the nearest store being a 45-minute drive away. It was very different from my life in Montreal.

My aunt and uncle left Ohio in a flash, eager to come grab me and take me home with them. I was dreading it.

The next night, their pickup truck bounced into our driveway.

"Ready to go?" Uncle Fuzz asked, his voice booming and his curly black hair almost brushing the ceiling. "We'd love to stay the night, but we have to get back for Anika's quilting lessons."

"Yeah," I muttered. His broad hands reached out for the handle of my suitcase.

"No, sir. I got this," I replied, trying to sound brave.

---

surreal: *unreal; dreamlike*

In the truck, Uncle Fuzz drove as we left the bright lights of Montreal behind and entered the darkness of the country. I sat there and wondered, *What the heck will I do in Ohio?*

"Now, you're probably a little upset. That's all right," my Aunt Frederica said. "But you'll only be here for the summer, and there are a lot of fun things to do. You could go with Anika to her quilting lessons," she said. "Isn't that right, Peaches?"

What do quilting and graffiti have in common? How are they different?

"Of course. I'm sure she'd love the company," said Uncle Fuzz.

Peaches? Quilting? I definitely wasn't in the city anymore! Just when I thought things couldn't get any worse, Uncle Fuzz turned up the country music on the radio — and I hated country music. I grabbed my iPod from my backpack, but no matter how loud I turned it, I could still hear the country music from the truck's speakers. This was going to be a long trip.

I must have hit my head or something because I passed out. I woke up and I heard a dog barking.

After a while, I just couldn't take it anymore.

"I need to go to the bathroom," I said.

Aunt Frederica turned to glare at me with her lips pursed. Right then, she looked just like my mom when my mom got angry at me for forgetting to say "please."

"Sorry, ma'am. May I please get out to use the bathroom? I really have to go … please," I said in my sweetest voice. *Ugh! They're killing me*.

Uncle Fuzz stopped the vehicle by the side of the road. I looked beside the gravel shoulder and saw the ditch beside it. It was so dark that I couldn't see the bottom of the ditch. I wondered if there were any snakes or leeches down there.

"Go on now, go to the bathroom," Uncle Fuzz said in his stern voice. I could not believe that they were trying to make me go to the bathroom there.

Aunt Frederica chimed in, "Listen to your uncle, Dwayne."

As I got out of the truck, I stepped on some loose gravel, which made me lose my balance and caused my feet to fly out from under me. I fell down the hill, landing in a big puddle at the bottom of the ditch. I must have hit my head or something because I passed out. I woke up and I heard a dog barking. At first, I thought it was barking in the distance, but it seemed to get closer and closer. My heart started racing — I was terrified of dogs. Out of nowhere, I felt someone's breath on my forehead.

"Yeah, he's still breathing, Sandy," a girl whispered. "You'd better get up and come with us before those hounds start following your trail. I'm Niki. Follow me."

Two arms pulled me up, and we immediately began to run. The ground felt hard and sharp under my feet — that's when I noticed that I didn't have any shoes on. I ran just the same, as the barking faded behind us.

I looked behind me, searching for a sign of my uncle's truck, but the road was empty. *Where did the truck go? It couldn't just vanish*. I ran for quite a while, following these two strangers, until we finally ended up at a log house. It was pitch-black inside, except for a single light in the window. There was a torn quilt with uneven patterns on it hanging on the front porch.

"Don't fret, you're safe now," Niki said to me as we approached the house. In the dim light, she looked almost like my cousin Anika, except her hair was pulled back and covered in a red scarf.

Niki explained that the house was owned by an elderly Black couple named Mr. and Mrs. Schmidt. They didn't have children of their own, and were too elderly to make the journey north, Niki continued to explain. "They're real nice folks, and are letting us stay at this station until we're ready to keep heading north."

Life was hard for enslaved people who attempted to escape to freedom through the Underground Railroad. They faced little food, long walks, dangerous conditions, and the constant risk of being caught.

"We're all alone, ever since our Ma was sold off and Pa died on the way to freedom, so we appreciate the Schmidts helping us out," Sandy said. My head was reeling. *What was happening? Where was I?* I started missing my mom. I think Niki noticed the change in my mood.

"Don't worry," she said. "We've helped plenty of people who've escaped from slavery in the south. You're on the Underground Railroad now. You're in good hands."

We didn't have any time to rest, and I didn't have any time to think about what Niki and Sandy had just told me. There was apparently a lot of work that needed to be done. We went out to meet Mr. Schmidt.

"We've helped plenty of people who've escaped from slavery in the south. You're on the Underground Railroad now."

"Come now, you can help chop some wood. Everyone's got to pull their weight," Mr. Schmidt said and handed me the handle of an axe. *This is unbelievable! An escaped slave? Travelling on the Underground Railroad? That's history!* My mind was still reeling.

I didn't know what was happening, but I could tell from Mr. Schmidt's expression he was serious about me helping out. Sandy chopped quickly and with lots of power. He made it look so easy. I tried to copy their movements, but I fumbled a bit and ended up dropping the axe.

"You ain't ever done this down south?" Sandy asked me.

Down south? I thought. *Is this a joke? They must really think I'm escaping from slavery.*

"No, but I learn quick," I managed to say under my heaving breath. It was more for myself than to Sandy. Who was I kidding?

"What did you do down south then?"

Remembering the graffiti that got me into this whole mess, I said, "Artist. I'm a painter."

"Oh," Sandy said.

"It's best you go inside then," Mr. Schmidt said with a snicker. "Mrs. Schmidt and Niki are working on another quilt. They could use an artist. The last quilt they designed came out crooked. Almost led the people back to the south." Both he and Sandy laughed, but their laughter was tinged with sadness.

I went back to the cabin, and the table was a mess. It was covered with piles of fabric, some plaid, some plain, and some with polka dots.

"I think we have to add this piece here," said Mrs. Schmidt.

"Are you sure, Mrs. Schmidt?" Niki scrunched up her nose, looking at her. When the old woman nodded, Niki bent over to pin the pieces together.

I walked over to Niki. "Can I see that?" I asked. I looked over the filmy white paper where patterns were drawn. Next to each pattern were a few words, such as "log cabin," "crossroads," and "star." *Are these meant to be instructions?* I wondered. There were quite a few patterns, too.

"You can read?" Niki asked. "I always wanted to read, but I was never allowed to learn how."

After I looked over the instructions, I knew exactly what to do.

"Niki, can you tell me the directions to get to freedom? And I'll tell you what quilt patterns you need."

"Okay," Niki agreed.

On the back of the paper, I sketched a quilt pattern based on Niki's instructions. I included all of the symbols to lead people to freedom. First, a crossroads. Then, the log cabin so that folks would know they were getting close to somewhere safe. Then, a star in the middle of the quilt. Beside it were blue lines that meant they would need to cross a river.

"How's that, Niki?" I asked her.

"Gee," she said. "You really are a true artist. Thanks, uh … I never did catch your name."

"Oh, it's Dwayne," I said.

"Dwayne?" she asked.

"Dwayne!"

All of a sudden, I awoke to Aunt Frederica standing over me, holding a cold cloth to my forehead. Bright lights shone in my eyes, and I had to squint.

"What happened? Where am I?" I stuttered.

"You're in a hospital," replied Aunt Frederica. "We thought you were a goner. You just couldn't wait to go to the bathroom, huh? I thought you were trying to run away," she said in a huff that had tiny notes of relief in it.

I wanted to answer but couldn't. I had no idea what had happened. Did I dream it all? My feet were sore and my body was tired. Maybe from running so much? It all felt so real.

I fell back asleep shortly after, thinking about the quilt patterns, and hoping they were real. I hoped that I really had helped lead people to freedom.

My feet were still sore and my body was tired. Maybe from running so much? It all felt so real.

CONNECT IT

Writer and activist Feminista Jones called Twitter the "underground railroad of activism." Do you agree that Twitter is a way to convey important messages? Write a short paragraph supporting your opinion.

SPEAKING OUT AND STEPPING UP

THINK ABOUT IT

An African proverb advises, "Until the lions have their own historians, the history of the hunt will always glorify the hunter." What do you think is meant by this proverb?

THE FAMOUS HIP-HOP group Public Enemy once said, "Rap is Black America's CNN." This meant that people of African descent used music to talk and provide information about their lives and experiences — stories that were often misrepresented in the mainstream media. As technology develops, we have more and more ways of communicating with one another. Traditional media roles are changing. The media is no longer just a reflection of our world, but a tool that anyone can use to bring about positive change. Read these quotations from a variety of media professionals to discover a range of views on media representation, hashtag activism, and more.

Reginald Hudlin

"Everyone knows diversity is good. We want Black superheroes, we want female superheroes, we want Latino superheroes. That makes things better. And they don't have to be sidekicks or buddies, they can be rock stars themselves."

— Reginald Hudlin, American writer, director, and producer

"Black culture is hugely influential in North America and we need to give it the space it deserves."

— Donna Bailey Nurse, Toronto-based literary journalist

"One thing, which seems very small, is not supporting shops and designers that aren't diverse in their advertising or runway shows. Money talks, whoever's holding it."

— Lana Ogilvie, Canadian fashion model

Lana Ogilvie

Tony Young

"I had a lot of memorable years on MuchMusic .... Right up until I left, I was blessed to have produced and hosted a show that played Black music and reflected our culture in a positive light."

— Tony "Master T" Young, English-born, Canadian-raised TV and radio personality

"I think [Black Twitter] became such a thing because via Twitter, a previously silenced group now has the opportunity to broadcast their thoughts and voices themselves, without having to go through a middle man that may or may not give them the stage."

— Tracy Clayton, American journalist

"As a kid growing up, the Black characters that I saw on television were never something I could relate to. They weren't role models. I had a dream as a kid to become an actor and tell stories for people that looked like they were from the Caribbean."

— Selwyn Jacob, Canadian documentary filmmaker

Selwyn Jacob

"Black Twitter is part cultural force, cudgel, entertainment, and refuge. It is its own society within Twitter, replete with inside jokes, slang, and rules, centred on the interests of young Blacks online — almost a quarter of all Black Internet users are on Twitter."

— Soraya Nadia McDonald, journalist

"Our politicians, media — none of them looked like me. And yet I was also being given this rhetoric of 'oh, we're so multicultural, we're so diverse,' but I didn't see it. And it was around the fifth grade that I started to ask questions. Why am I not seeing myself reflected?"

— Rita Nketiah, Canadian writer of Ghanaian descent

rhetoric: *language that is designed to persuade but is not sincere*

Karlene Nation with her son

"I am proud of my responsibilities in terms of helping CTV to broaden our coverage of diverse communities, and to widen the representation of people, from all backgrounds, in our local and national news programming."

— Karlene Nation, Jamaican-born journalist turned political candidate (came to Toronto from Jamaica in 1976)

"When choosing which social network to spend your time with, consider what your goals are, how much time you want to spend online, and just how much you plan to engage with the tools."

— Juliette Powell, American-born Canadian author, TV host, and producer

"Black Media has an amazing opportunity (and responsibility) to serve its readers by providing more in-depth science news coverage. … It's time for the conversation to change, to mature. It's past time for the Black community to have very serious conversations about science which includes access to quality education so that young people can pursue these 21st century career opportunities."

— Danielle N. Lee, American biologist and blogger for *Scientific American*

In 2014, after 18-year-old Michael Brown was killed by a police officer in Ferguson, Missouri, United States, many Black people took to Twitter to protest. They used the hashtag #IfIWasGunnedDown to show how mainstream media tends to use images of Black people who have been killed that show them in a stereotypical light. For example, the media is less likely to show a Black teen in a graduation cap than a White teen. This perpetuates the harmful stereotype that Black teens are not educated.

"Following the events in Ferguson, Mo., that resulted in the death of Michael Brown, Black Twitter forced the mainstream media to pay attention by rallying behind the power of the hashtag. Emerging from that tragedy … were a series of movements that showcased just how powerful African American voices can be when they unite for a common cause."

— Trent Jones, American online media professional

CONNECT IT

#BlackLivesMatter is a hashtag that began in 2012 with the killing of Trayvon Martin, an unarmed 17-year-old, and has been used to protest subsequent killings of Black people. It says that the life of a Black person is just as valuable as everyone else's. Using this hashtag, create a series of social media posts to share your views.

Mom Gives Barbies a Multicultural Makeover

THINK ABOUT IT

What do you think the term "dominant culture" means? In a small group, discuss why the media often portrays only the dominant culture instead of diverse cultures.

SAMANTHA CRAGGS
CBC NEWS
14 APRIL 2014

GROWING UP, KIDS learn a lot of things from different types of media. But media don't always reflect the lives and realities of all people. One mother in Hamilton, Ontario, decided to change that. She designed diverse Barbie dolls to better reflect her daughter's identity. She shared her creations using social media. Read all about her in this online article.

Queen Cee Robinson with one of the dolls she customized

It was about a year ago that Queen Cee Robinson tried to find a doll that resembled her daughter and had a revelation — there weren't any.

Robinson had seen Black Barbie dolls before. Most of them wore bikinis, and they all had long, straight hair. And they all looked the same — sidekicks to the White dolls, or as Robinson describes them, "Barbie's token Black friend."

That started Robinson down a path of giving dolls makeovers — providing them with loose curls and dreadlocks, hijabs and sarongs, all in the name of giving little girls a realistic image of themselves.

Growing up, Robinson never played with dolls much herself. But the more she looked for a doll for her six-year-old daughter, the more upset she became at the limited choices.

"It's always been a focus as a young Black girl growing up," she said. "You want to see something that looks like you, and that's very rare and scarce to find in media and promotions and toys."

Robinson searched online and eventually found the Mattel line of So In Style dolls. Most of them still had straight hair, but they deviated from the majority of the homogeneous Barbies in stores.

The So In Style dolls were a line of Black Barbie dolls, with varying skin tones.

Through more research, she found that the So In Style dolls weren't carried in Canada, and even more, Mattel had discontinued the line.

Robinson bought up as many of the remaining dolls as she could and "reimaged" one for her daughter, giving it curly hair and a new outfit. She posted pictures [on] social media, and word spread. Since then, she's done about a dozen custom dolls for people who have contacted her. She charges for the time and materials. Eventually, she'd like to create a proper business making dolls of various ethnicities.

How is social media helpful for getting more diverse voices heard?

She also uses the dolls for her non-profit Bee-You-Tiful Girls Club, which gives girls creative outlets to express their identities. In February, she held a Just Like Me workshop at the Hamilton YWCA where girls used the dolls to create characters and tell stories aimed at empowering and inspiring them.

---

scarce: *hard to find*
deviated: *did something different from what is usual*
homogeneous: *similar*

One of Queen Cee Robinson's customized dolls

Practising on Monster High dolls

Reimaging dolls has been a trial-and-error process for Robinson, a Hamilton-based singer/songwriter, and mom of five.

She learned to remove face paint by practising on Monster High dolls. She designs and makes the clothes herself. She uses mohair for the hair, stitching it in strand by strand and then styling it to her customer's specifications.

"The hair is a major thing because that is what sets my daughter apart from someone of another race," she said. "It's distinct to her and it's beautiful, and I want her to be comfortable with that."

Women of all ages have taken notice. Robinson has taken orders from adults who want the dolls as keepsakes. She hears from moms of all ethnic backgrounds who want more diverse doll collections for their children.

It's important that children grow up with positive toys and images that reflect them, said Susan Fast, a McMaster University professor of cultural studies. That extends to what TV shows and movies they see.

"If children only have an image of what the dominant culture looks like, that's problematic," she said.

Monster High dolls are characters who are children of famous monsters, including Dracula and Frankenstein. Monster High dolls and toys are also featured in TV specials, video games, and other products.

In this context, the dominant culture in Canada refers to the widespread representation and influence of White people.

Creating positive images

"It goes beyond just having minorities represented in some way, which I do think is really important, to how they get represented."

White [people] tend to be seen as having purchasing power, Fast said, and toy makers want to reach the largest markets possible. White [people] also tend to be in senior positions at toy companies, she said, so they're the ones controlling the images.

If interest in Robinson's creations is any indication, there is a market for dolls showing different ethnic backgrounds. Canada is increasingly multicultural, Robinson said, and toys should reflect that.

"What is reflected to you as a child is ultimately what you're going to grow up to think like, or ultimately what you're going to become," she said. "We have to allow them to create positive images of themselves when they're young."

Why is it important to have people from diverse backgrounds control how they are represented in media? What is wrong with toy makers creating products only for White people?

---

mohair: *either cotton or wool yarn made from goat hair*

CONNECT IT

Using the Internet, research some of the dolls Robinson has made. Create a character profile for one doll, and write a story about it that is empowering. Illustrate your story, and present it to a partner.

WHAT DO YOU WANT TO BE?

THINK ABOUT IT

With a partner, discuss a media-related career that interests you. What is it that interests you about this career?

MANY PEOPLE WORK at jobs that are related to the media. These jobs require diverse skills. Read the following fact cards to learn about just a few of these interesting jobs.

Travel and Lifestyle Writer

Heather Greenwood Davis

JUST LIKE: Heather Greenwood Davis

CURRENT CITY: Toronto, Ontario

WHY DID YOU CHOOSE THIS CAREER? I've wanted to be a writer since elementary school. I loved to read and saw the power of stories (fiction or non-fiction) to impact people's lives. I wanted to help tell those stories.

WHAT IS YOUR JOB DESCRIPTION? I'm a freelance writer. It means that I'm self-employed and hire my writing skills out on a contract basis to organizations (newspapers, magazines, brands, other writers …) to use my skills to help tell stories. …

WHAT DO YOU LIKE MOST ABOUT YOUR CAREER? The flexibility. My hours are my own to set. That's helpful given that my kids are young and I like being involved with their school day. It also allows me to travel around the world — a passion of mine — and tell the stories I feel passionate about.

WHAT ARE THE CHALLENGES OF YOUR CAREER? I only earn when I'm writing. That means true vacations can be hard to come by. Luckily, I love what I do.

Character Designer

JUST LIKE: Sean L. Moore

CURRENT CITY: Vancouver, British Columbia

WHY DID YOU CHOOSE THIS CAREER? I discovered very early (grade 7) the path I wanted to take. The idea of creating characters and bringing them to life was always fascinating to me. Whether it was through puppetry, Claymation, or animation, that was what I was going to do.

WHAT IS YOUR JOB DESCRIPTION? As a character designer, I create any new characters required for an episode on any given show … It's almost like building actors from scratch. … In character design, creating a personality and telling a story through a single image … are the most important aspects of the job. …

WHAT DO YOU LIKE MOST ABOUT YOUR CAREER? As much as things are revised and/or edited by the end of a project, ultimately my job allows me to create without limitation. I get to continue to use my imagination when so many people my age and much younger have lost theirs.

WHAT ARE THE CHALLENGES OF YOUR CAREER? The biggest challenge of my career is the same thing that keeps me motivated — the fear of running out of ideas. Everything is derived from something else, especially in the arts. But the fact that creative people, young or old, continue to find ways to innovate is very inspiring. …

---

derived: *based on; obtained from*

Sean L. Moore

Character design begins with sketches on paper or on a computer. They are then developed further using a computer.

Filmmaker

JUST LIKE: Stefan Verna

CURRENT CITY: Montreal, Quebec, but originally from Kinshasa, Democratic Republic of the Congo

WHY DID YOU CHOOSE THIS CAREER? [I wanted to] be an agent of change in my community and in the world at large, but also because I love telling stories that encourage us to grow.

WHAT IS YOUR JOB DESCRIPTION? [It includes writing,] filming, editing, and promoting films.

WHAT DO YOU LIKE MOST ABOUT YOUR CAREER? Cinema allows me to share my passion for the many aspects of life. I love the collaborative aspect of making films.

WHAT ARE THE CHALLENGES OF YOUR CAREER? [Challenges are] finding funding for my independent film projects, [and] balancing my time between commercial and personal films

Stefan Verna

Sahle Robinson

Storyboard samples done by Sahle Robinson

Story Artist

JUST LIKE: Sahle Robinson

CURRENT CITY: Toronto, Ontario

WHY DID YOU CHOOSE THIS CAREER? I decided to become a story artist because the job is very enjoyable and esteemed, plus it's the position in animation that my talents were best suited for.

WHAT IS YOUR JOB DESCRIPTION? Storyboard artists (story artists or "storyboarders") draw scenes by hand or computer … The storyboard presents the "action" in a series of scenes, allowing for evaluation of the project before beginning production, as well as providing direction during production. In addition to TV and film, storyboard artists work in advertising, music videos, and the video game industry.

WHAT DO YOU LIKE MOST ABOUT YOUR CAREER? What I like most about my career is that … it's the one job that encompasses and influences all aspects of making a show (production). …

WHAT ARE THE CHALLENGES OF YOUR CAREER? The most challenging aspect of my career is the amount of time and brain energy it takes up. Also, when you do art for money, it tends to distort your love for the art!

What do you think Robinson means by saying making art for money distorts your love for art? Do you agree or disagree?

esteemed: *respected*
distort: *change from the original*

Graphic Designer

Colanthony Humphrey

JUST LIKE: Colanthony Humphrey

CURRENT CITY: Toronto, Ontario

WHY DID YOU CHOOSE THIS CAREER? I really didn't, actually. With an entrepreneurial mind, you acquire different skills that you wilfully turn into a commodity. Always doing art, being a designer was a natural progression and, actually, the step before art director, which is what I'm en route toward.

WHAT IS YOUR JOB DESCRIPTION? Working directly with an art director and account team to visually achieve a client's goal as it relates to graphic design. ... The graphic designer is the last person to touch a piece before it goes to print. ...

WHAT DO YOU LIKE MOST ABOUT YOUR CAREER? I like the opportunity to create. ... Advertising allows and thrives on creativity, though.

WHAT ARE THE CHALLENGES OF YOUR CAREER? Extremely long hours and short deadlines make some days harder than others.

---

entrepreneurial: *business savvy*
commodity: *valuable or useful thing*

Chief Curator

JUST LIKE: Michelle Jacques

CURRENT CITY: Victoria, British Columbia

WHY DID YOU CHOOSE THIS CAREER? I became a curator because I was interested in working directly with artists.

WHAT IS YOUR JOB DESCRIPTION? My primary responsibilities are the development, care, and presentation of the collection, and producing exhibitions. In my current role as a chief curator, I am also responsible for setting the artistic vision of the art gallery at which I work, and I manage the curatorial and education departments.

WHAT DO YOU LIKE MOST ABOUT YOUR CAREER? The thing I like most about my career is the variety involved in what I do. I work directly with art and artists, I write, [I] give tours and lectures, [and I] do administration like budgeting, scheduling, grant writing, and planning. I also get to work with people throughout my institution and the community.

WHAT ARE THE CHALLENGES OF YOUR CAREER? The biggest challenge of being a curator is the same as what I like most about it. The variety means that you need a wide range of skills, have to be able to multitask, and most importantly, are a good communicator and collaborator.

Michelle Jacques

CONNECT IT

Find out more about the career you chose for the Think About It activity. Research information to write a fact card about this career. Be sure to include a job description, what you think you would like best about this career, and what you think the biggest challenge would be.

These T.V. Men and Women

THINK ABOUT IT

What is your favourite program to watch on television? Is there a person or a character you connect with? Think about why this program or person has an impact on you.

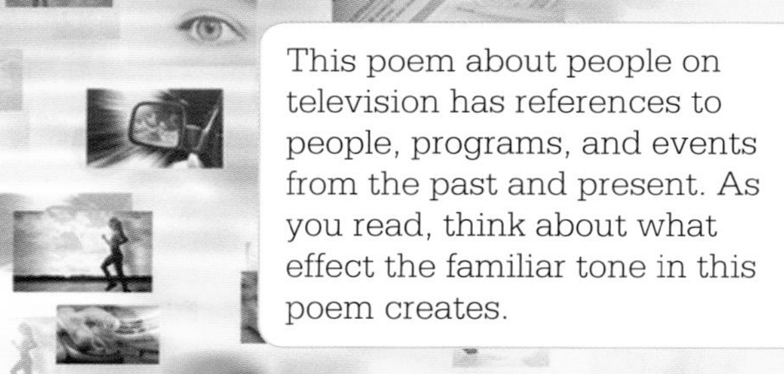

This poem about people on television has references to people, programs, and events from the past and present. As you read, think about what effect the familiar tone in this poem creates.

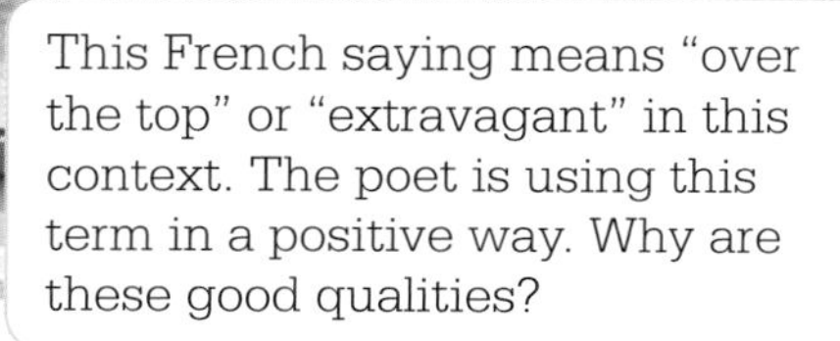

This French saying means "over the top" or "extravagant" in this context. The poet is using this term in a positive way. Why are these good qualities?

ABOUT THE POET

Maxine Tynes was a celebrated poet and teacher. She was born in Dartmouth, Nova Scotia, and later became the first African Canadian woman to sit on Dalhousie University's Board of Governors. She wrote many poetry collections, and she received the Milton Acorn People's Poet Award.

BY MAXINE TYNES

These t.v. men and women people my life
with names like David and Joan,
Norm and Barbara and Oprah
and Johnny;
they smile and turn and move
in animated, air-brushed perfection
the glint and pouf and style
of each hair,
the sweep of a skirt,
taut and tight leg and neck and back,
the perfect eye
hands flutter bejewelled and multihued
by Revlon, Max Factor, L'Oreal, Maybelline.

These t.v. men and women;
passion and pain are fleeting in their perfection,
trauma is a 30-60-90-minute
rise and fall and test of
emotional reality/non-reality,
theirs and mine.
At arms-length, I am rich
and svelte and amorous and
tres, tres outre,
full of scandal and intrigue.
This vicarious multi-life we all
live and share with them
via channels and cable and
to-pay-or-not-to pay t.v.

---

taut: *firm*
multihued: *multicoloured*
svelte: *graceful; elegant*
amorous: *inclined to romantic love*
vicarious: *experienced through the imagination*

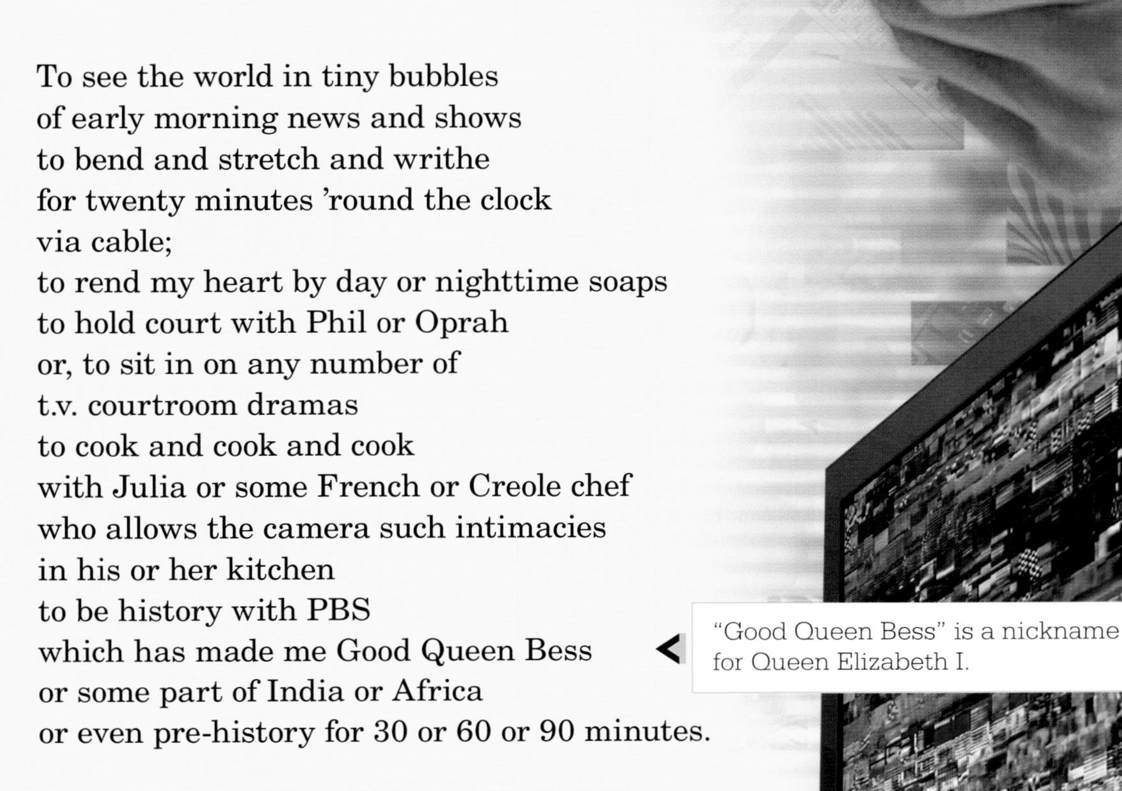

To see the world in tiny bubbles
of early morning news and shows
to bend and stretch and writhe
for twenty minutes 'round the clock
via cable;
to rend my heart by day or nighttime soaps
to hold court with Phil or Oprah
or, to sit in on any number of
t.v. courtroom dramas
to cook and cook and cook
with Julia or some French or Creole chef
who allows the camera such intimacies
in his or her kitchen
to be history with PBS
which has made me Good Queen Bess
or some part of India or Africa
or even pre-history for 30 or 60 or 90 minutes.

"Good Queen Bess" is a nickname for Queen Elizabeth I.

I have seen the holocaust from many angles.

I examine South Africa again and again.
I, with billions of world viewers
exploded with the fated 1986 space shuttle
over and over and over again
as it disintegrated in a t.v. sky.

People are often brought together during tragedies by watching them on television. What other ways can television create a sense of community?

Oh, television
you chart and record my pre and current
and post history;
you are relentless in pursuit of my every
sleeping and waking hour.
Your people tattoo themselves on my life
in this surreal and glamorous
flicker and dim;
TV,
you orchestrate my life.

CONNECT IT

Reread this poem. Create a T-chart that lists the positive and negative things that television brings to the speaker's life. Using this chart, write a paragraph about whether the speaker's relationship with television is mostly positive or mostly negative.

MEDIA WHIZ KIDS

THINK ABOUT IT

If you could be the first to invent something amazing, what would it be?

YOUNG PEOPLE ARE using media and technology in new and innovative ways. Read the following profiles to learn about a few media whiz kids.

Kelvin Doe

KELVIN DOE

Freetown, Sierra Leone, Africa

Kelvin Doe had a dream — to be a DJ and host his own radio show. From the age of 10, Kelvin gathered scrap metal and built transmitters, amplifiers, radios, generators, and batteries — all the things he needed to make his own radio station. From his own radio station, Kelvin played music and news under the name DJ Focus. In 2013, when he was 16, Kelvin signed a $100,000 solar project agreement with Canadian high-speed service provider Sierra WiFi. Kelvin was the youngest visiting practitioner at the Massachusetts Institute of Technology (MIT).

WondaGurl

WONDAGURL, A.K.A. EBONY OSHUNRINDE

Brampton, Ontario, Canada

Born in 1996, WondaGurl is making a name for herself in hip hop. At the age of nine, she began making beats. She was first noticed when she was 14 and made it to the quarter-final of the 2011 Battle of the Beat Makers in Toronto, Ontario. Producer Boi-1da noticed her and began mentoring her. Since then, she won the 2012 Battle of the Beat Makers. She has also produced a track that was used on Jay Z's album *Magna Carta… Holy Grail*, which was up for the Best Rap Album at the 56th Grammy Awards. How did she get started? WondaGurl is self-taught. She watched videos on YouTube and taught herself to use FL Studio, software that lets anyone produce music.

JAYLEN BLEDSOE

Hazelwood, Missouri, United States

Jaylen Bledsoe started his own company, Bledsoe Technologies, before he even turned 16 years old. The company specializes in web design and other information technology (IT) services. Jaylen is an investor, technology consultant, youth rights advocate, and motivational speaker. Of his work as a youth rights advocate, Jaylen said, "Often in society, we expect that our youth care more about making and spending a dollar on themselves, rather than helping someone else out." Jaylen helps people realize that kids can also make a difference. Jaylen received the President's Award for Educational Excellence from President Obama in 2009. When he graduates from high school, Jaylen plans to attend Harvard University.

Jaylen Bledsoe

CONNECT IT

With a group, brainstorm a new media product or service you think would be useful in your lives. What skills would you need to invent this product or service? What marketing tools would you use to convince others that they need to buy it?

PEOPLE OF ALL SHADES

THINK ABOUT IT

With a classmate, look through a magazine and make a note of the people featured in it. Are people of different races featured in the pages and the ads?

HAVE YOU EVER thought about where your ideas of beauty come from? Are they from your parents? Your friends? Celebrities? Chances are the media plays a large role in shaping your opinions about what is beautiful and what is not. Read this report to learn more about how the media's portrayal of skin colour can have an effect on self-esteem.

In a perfect world, the media would provide an accurate reflection of society. Sometimes, however, certain groups of people, including racialized groups, are not shown in the media or are shown unfairly. People from racialized groups are often shown as stereotypes, or only people who look a certain way, such as those with light skin, are shown. When people compare themselves to these images, they often feel inadequate.

Lupita Nyong'o, the Oscar-winning actress and fashion's "It" girl of 2014, has spoken before about how seeing only pale skin on television made her feel "unbeautiful." Nyong'o said she was teased for having dark skin and that it wasn't until she saw supermodel Alek Wek that she began to see her skin shade as beautiful. Because of Wek, Nyong'o, who was born in Mexico to Kenyan parents, saw herself reflected in the media.

What Nyong'o experienced was feeling inadequate because of shadeism. Shadeism, also known as colourism, is discrimination based on skin tone or shade. Some forms of shadeism originated from the way slave owners would give preferential treatment to the lighter-skinned children of Black enslaved women. The children were the result of enslaved women being forced to have sex with slave owners.

Lupita Nyong'o poses after winning the Oscar for Best Supporting Actress for her role in the 2013 film 12 Years A Slave.

There are two kinds of shadeism. Interracial shadeism happens when one racial group discriminates against another based on skin colour. Intraracial shadeism happens when members of a racial group discriminate against other members of the same group based on skin colour.

In 2014, Nyong'o partnered with Lancôme, a cosmetics company, to become the company's first-ever Black ambassador. Nyong'o was hired to promote a new brand of foundation that was available in 28 different tones for all shades of skin. Some women of African descent say they feel disconnected from products that advertise using only light-skinned women. By pairing with Nyong'o and having its makeup available in 28 shades, Lancôme is celebrating the diversity of its audience instead of ignoring it.

DOCUMENTING SHADEISM

Lupita Nyong'o isn't the only one changing the face of the media today. In 2010, students at Toronto's Ryerson University teamed up to produce a documentary called *Shadeism*. The documentary was created to help women of colour talk about the issue of shadeism. Muna Ali, the documentary's executive producer, said, "It harms our identity. It harms our self-esteem." *Shadeism* is now being made into a full-length documentary for release in 2014.

Nayani Thiyagarajah, the director, wanted to make the documentary because she saw how shadeism was affecting her three-year-old niece, who already thought that her darker skin tone kept her from being beautiful.

The documentary also mentions the growing industry of skin-bleaching products and the advertisements that tell consumers that fairer or lighter skin is more beautiful. Some magazines and advertisers have been accused of lightening the skin tone of African American celebrities, including Gabourey Sidibe, Halle Berry, and Beyoncé. According to the market research firm *Global Industry Analysts*, the market for skin-bleaching products is expected to reach US$10 billion by 2015. A *Toronto Star* article about shadeism also noted that some dark-skinned women use filters on Instagram in order to appear lighter. With Instagram, people can control how they are presented to the world — and many still choose to follow what they see in advertisements or other forms of mass media.

The media is always telling us a story. Ali says, "Just think about what you learn from somebody telling you that story; think about the power invested in that story." People like Thiyagarajah and Ali are sharing their story to raise awareness about the issue of shadeism, not only in the media, but also in our society.

Gabourey Sidibe

CONNECT IT

Imagine you are marketing a product of your choice. Create an advertisement for this product that promotes diversity. You can choose any medium you like. For example, you could design a poster or write an advertorial (advertisement that looks like an editorial or article). Consider your audience and the story you want your advertisement to tell. Share it with your class.

AFROFU

THINK ABOUT IT

Why might science fiction and futuristic music and art be good places to address social issues?

ALIEN ABDUCTIONS? SPACESHIPS? New planets? Androids? How does this connect to the experiences of people of African descent? They're all part of Afrofuturism. Media writer Mark Dery created the term "Afrofuturism" in the early 1990s. He used it to describe science fiction that talks about the experiences of Black people.

Science fiction is often used by artists to talk about social problems and to imagine new possibilities. Black artists have used Afrofuturism to explore different aspects of Black history and culture, including exploitation and freedom. For example, in Afrofuturism, fighting with an alien can be a metaphor for racism. Alien abductions can be a metaphor for the transatlantic slave trade. The transatlantic slave trade kidnapped Africans, bringing them to new lands with different languages. That doesn't sound so different from an alien abduction after all.

Read the following profiles to learn about different artists using Afrofuturism.

Androids: *robots that appear human*

Janelle Monáe

Janelle Monáe
Medium: Music

In 2007, Janelle Monáe recorded *Metropolis*, an album that follows the story of an android named Cindi Mayweather. Monáe's track "Many Moons" reimagines the classic 1927 science fiction film *Metropolis*, which was acclaimed for its criticism of the exploitation of workers.

A short film released with "Many Moons" shows androids being auctioned off. The androids have names associated with people from racialized backgrounds. The video ends with a quote from Cindi Mayweather: "I imagined many moons in the sky lighting the way to freedom."

What might this quote say about the relationship between science fiction and the fight for freedom?

TURISM

Nalo Hopkinson
Medium: Novels and Short Stories

Nalo Hopkinson combines Caribbean traditions and folklore with futuristic worlds in her stories. Hopkinson has spent most of her life moving between Toronto and the Caribbean, where she was born. Her second novel, *Midnight Robber*, takes place on the imaginary planets Toussaint and New Half-Way Tree. Toussaint, named after a leader of the Haitian Revolution in the late 18th century, is a colonized planet meant to symbolize the Caribbean. New Half-Way Tree is a planet where criminals are exiled to and where Caribbean folklore comes to life. *Midnight Robber* is about how colonialism affects people and places.

colonialism: *exploitation of land or people by people with more power or wealth*

Nalo Hopkinson

Wanuri Kahiu
Medium: Film

Wanuri Kahiu

Wanuri Kahiu is a Kenyan filmmaker whose first science fiction film was a short film called *Pumzi*. *Pumzi* is set in the future, where a war over global water shortages has forced people to live underground. The main character is a botanist who is trying to locate soil to nurture plants back to life. When Kahiu was doing research for *Pumzi*, she looked at both 1950s science fiction films and traditional African artwork. She wanted African art and tapestries to form the background for *Pumzi*.

The filmmaking industry in Kenya is still growing, but Kahiu has hopes for it. Of Kenya, she said, "It has stories, it has people … These things are beginning to make a difference, so people want to hear more about Africa, an authentic Africa." Focus Features created the Africa First short film program, and *Pumzi* was made with the grant money from this program.

Still from Pumzi

George Clinton
Medium: Music

In 1975, George Clinton's funk group Parliament wrote songs about sending Black people into outer space. The group released the concept album *Mothership Connection*, which combined science fiction and civil rights. During one of the band's concerts, a spaceship descended from the ceiling. Kevin Strait, a specialist at the National Museum of African American History and Culture, defined this stage prop as a symbol of the "liberating power of music." For Clinton, envisioning stories about outer space was a way for people of African descent to talk about and free themselves from racism, poverty, and other social issues.

George Clinton and Parliament performing live at the HMV Ritz, in Manchester, England, on 1 December 2011

Komi Olaf
Medium: Visual Art

Komi Olaf

Komi Olaf is a Nigerian-born visual artist whose paintings explore African and Canadian identities. Olaf's painting *3014* depicts an African woman in the future, blending together technology and nature. According to Trend Hunter, the woman's name in the painting is Alkebulan, the name some refer to as the original name of Africa. According to Olaf, it is important to celebrate both where we are from and where we are going.

Why do you think it's important that Olaf may have named the woman in his painting Alkebulan? What does this say about how the past and the future are linked?

3014 *by Komi Olaf*

CONNECT IT

What do you think your community will be like in the future? Design a poster with words and images that reflect your thoughts. Summarize your ideas about what the future might be like to a classmate.

Index

A
activism, 18–19, 27–28
Adidas, 14, 16
advertisement, 7, 14, 17, 28, 36–37, 43
Afrofuturism, 44–47
Air Jordan, 14–17
Ali, Muna, 43
animation, 8, 10–13, 35–36, 38
artist, 16, 26–27, 37, 44, 47

B
Barbie, 31–32
Beyoncé, 43
Black Twitter, 29–30
Bledsoe, Jaylen, 41
broadcast, 13, 29
Brown, Michael, 30

C
cartoon, 10–13
character designer, 35
chief curator, 37
Clinton, George, 46
consumer, 15, 17, 43
customer, 17, 33

D
Dassler, Adolf "Adi," 14
diversity, 7, 13, 21, 28–34, 43
Doe, Kelvin, 40
dominant culture, 31, 33
Drake, 16

E
education, 4, 8, 11–13, 19, 21, 30, 37, 41
endorsement, 15–16

F
film, 8–9, 21, 36, 42, 44, 46
filmmaker, 8, 29, 36, 46

G
graffiti, 24, 26
graphic designer, 37

H
hashtag, 28, 30
hip hop, 16, 28, 41
Hopkinson, Nalo, 45

I
Instagram, 43
Internet, 4, 13, 29, 33

J
Jordan, Michael, 15–16

K
Kahiu, Wanuri, 46
Kami, 13
Kola, Vashtie, 17

M
magazine, 21, 34, 42–43
market, 12–13, 33, 43
marketing, 14–15, 41, 43
Mawa, Moses A., 21
Mawa, Patricia Bebia, 21
Monáe, Janelle, 44
music, 14, 16, 24, 28–29, 36, 40–41, 44, 46

N
newspaper, 14, 18–21, 34
Nike, 15–17
Nyong'o, Lupita, 42–43

O
Olaf, Komi, 47
Owens, Jesse, 14

P
Provincial Freeman, 18–21

Q
quilt, 22–27

R
racialized group, 6–7, 42, 44
radio, 24, 29, 40
rap, 16, 28, 41
representation, 6–7, 9, 28, 30, 33
Robinson, Queen Cee, 31–33
Run-DMC, 16

S
Shadd, Mary Ann, 18–21
shadeism, 42–43
Sidibe, Gabourey, 43
social media, 17, 30–32
stereotype, 4, 6–7, 30, 42
story artist, 36

T
technology, 28, 40–41, 47
television, 4, 6–7, 10–11, 13, 18, 21, 29–30, 33, 36, 38–39, 42
Thiyagarajah, Nayani, 43
Twitter, 22, 27, 29–30
Tynes, Maxine, 38

U
Underground Railroad, 19, 22, 25–27

V
video game, 4, 7, 33, 36

W
Waziri, Adamu, 10–13
WondaGurl, 41
writer, 12, 27–29, 33–34, 44

Acknowledgements

Craggs, Samantha. "Hamilton Mom Gives Barbies a Multicultural Makeover," from CBC News, 14 April 2014. Copyright © CBC 2014. All rights reserved. Reprinted with permission.

Ngoh, Helen. "Bino and Fino: a Nigerian Cartoon Brand for Kids," from *VC4Africa* blog, 26 March 2012. © 2014 *VC4Africa*. Permission courtesy of Venture Capital for Africa.

Tynes, Maxine. "These T.V. Men and Women," from *Woman Talking Woman*. Pottersfield Press, 1990. Permission courtesy of Pottersfield Press.

Photo Sources

Cover: [girl–Darren Baker; media icons–Peshkova] Shutterstock.com; **4:** wave–amalia19; controller–BonD80; media collage–Oleksiy Mark] Shutterstock.com; **6:** [background–winui; people icons–Leremy; tvs–Fernando Eusebio] Shutterstock.com; **7:** map–Filip Bjorkman/Shutterstock.com; **8:** background–winui; film–1nana1; camera–Anthonycz] Shutterstock.com; **9:** volunteers–lolakenyascreen.org; **10:** [wavy pattern–Aubord Dulac; pattern–marmarto] Shutterstock.com; Adamu Waziri–Ibraham Waziri / EUCL Studios; **11:** Fino–Ibraham Waziri / EUCL Studios; **12:** family–Ibraham Waziri / EUCL Studios; **13:** Zeena–Ibraham Waziri / EUCL Studios; Kami–Photo by John Barrett/Sesame Street/Getty Images; **14:** lights texture–javarman; cracked texture–isaravut; laces–Bayanova Svetlana; Jesse Owens–Owens3; Adidas shoe–copyright adidasAG/Studio Waldeck/adidas.com; **15:** laces–Bayanova Svetlana/Shutterstock.com; Michael Jordan–photo by Scott Cunningham/NBAE/Getty Images; Air Force 1–photo by Focus on Sport/Getty Images; **16:** Run-DMC–photo by Michael Ochs Archives/Getty Images; Drake–bukley/Shutterstock.com; **17:** Vashtie Kola–Photo by Johnny Nunez/WireImage/Getty Images; **18:** [background–Eky Studio; metal–siro46] Shutterstock.com; newspaper–Archives of Ontario; **19:** Mary Ann Shadd–Library and Archives Canada / C-029977; **20:** Shadd statue–courtesy of Kaitlin Tremblay; **21:** Patricia Bebia Mawa–US Embassy Canada; **22:** train background–Paul Matthew Photography; paint splats–Ghenadie; graffiti guy–morrison77; paint spray–Robles Designery] Shutterstock.com; **23:** police car–travis manley/Shutterstock.com; **24:** road–Sean Pavone/Shutterstock.com; **25:** boy–PathDoc; running silhouettes–KoQ Creative] Shutterstock.com; **26:** [boy–PathDoc; axe–Kletr] Shutterstock.com; **27:** quilt–Marinerock; spool–windu] Shutterstock.com; **28:** background–Carlos Caetano/Shutterstock.com; Reginald Hudlin–Featureflash/Shutterstock.com; Lana Ogilvie–ZUMAPRESS.com/Keystone Press; **29:** Tony Young–Photo by George Pimentel/WireImage/Getty Image; Selwyn Jacob–HEINZ RUCKEMANN/UPI/Newscom; **30:** Karlene Nation with her son–photo by Rick Madonik/Toronto Star via Getty Image; **31:** [background–scyther5; lines–wongwean] Shutterstock.com; Queen Cee Robinson–courtesy of Queen Cee Robinson; **32:** measuring tape–OlgaNik/Shutterstock.com; custom doll–courtesy of Queen Cee Robinson; **33:** scissors–OlgaNik/Shutterstock.com; **34:** Heather Greenwood Davis–CL Buchanan Photography; **35:** Sean L. Moore–courtesy of Sean L. Moore; kids–Gary John Norman/Glow Images; sketches–Oscar Chávez; **36:** film reel–Dinga/Shutterstock.com; Stefan Verna–courtesy of Stefan Verna; Sahle Robinson–courtesy of Sahle Robinson; storyboards–courtesy of Sahle Robinson; **37:** Colanthony Humphrey–courtesy of Colanthony Humphrey; Michelle Jacques– courtesy of Michelle Jacques; **38:** [halftone–vectoraart; image collage–Angela Waye] Shutterstock.com; **39:** hand holding remote–bikeriderlondon; tv–cobalt88] Shutterstock.com; **40:** tech background–kentoh/Shutterstock.com; Kelvin Doe–John Dalton; **41:** WondaGurl–Lucas Oleniuk / GetStock.com; Jaylen Bledsoe–courtesy of Jaylen Bledsoe; **42:** [Lupita Nyong'o–Helga Esteb; brick wall–Peshkova; geometric pattern–qushe] Shutterstock.com; **43:** filmstrip–teacept/Shutterstock.com; Gabourey Sidibe–Everett Collection/Shutterstock.com; **44:** space background–diversepixel; tech background–Adrian Grosu; halftone–Tusiy; Janelle Monáe–S_bukley] Shutterstock.com; **45:** Nalo Hopkinson–www.nalohopkinson.com; **46:** Wanuri Kahiu–Joshua Wanyama; sky background–Oxa/Shutterstock.com; Pumzi–NSPIRED MINORITY PICTURES/Album/Newscom; George Clinton & Parliament Funkadelic–ZUMAPRESS.com/Keystone Press; **47:** Komi Olaf–courtesy of Komi Olaf; 3014 by Komi Olaf–courtesy of Komi Olaf.